Visit Susie and her Animal Friends at
KidsKyngdom.com

For my First Born
Dominic
The Dominator

Thank you for showing us how to **Live a Joyous Life**

Thank you for supporting Kids Kyngdom

Get Your **FREE**
audio book
at
KidsKyngdom.com

Send Dr. Susie a Creative Message on Instagram

Instagram

I may pick You!

Thank You Kyngs!

KidsKyngdom.com

Hi Kyngs!

It's me. Susie!

My friend Bash and I are traveling

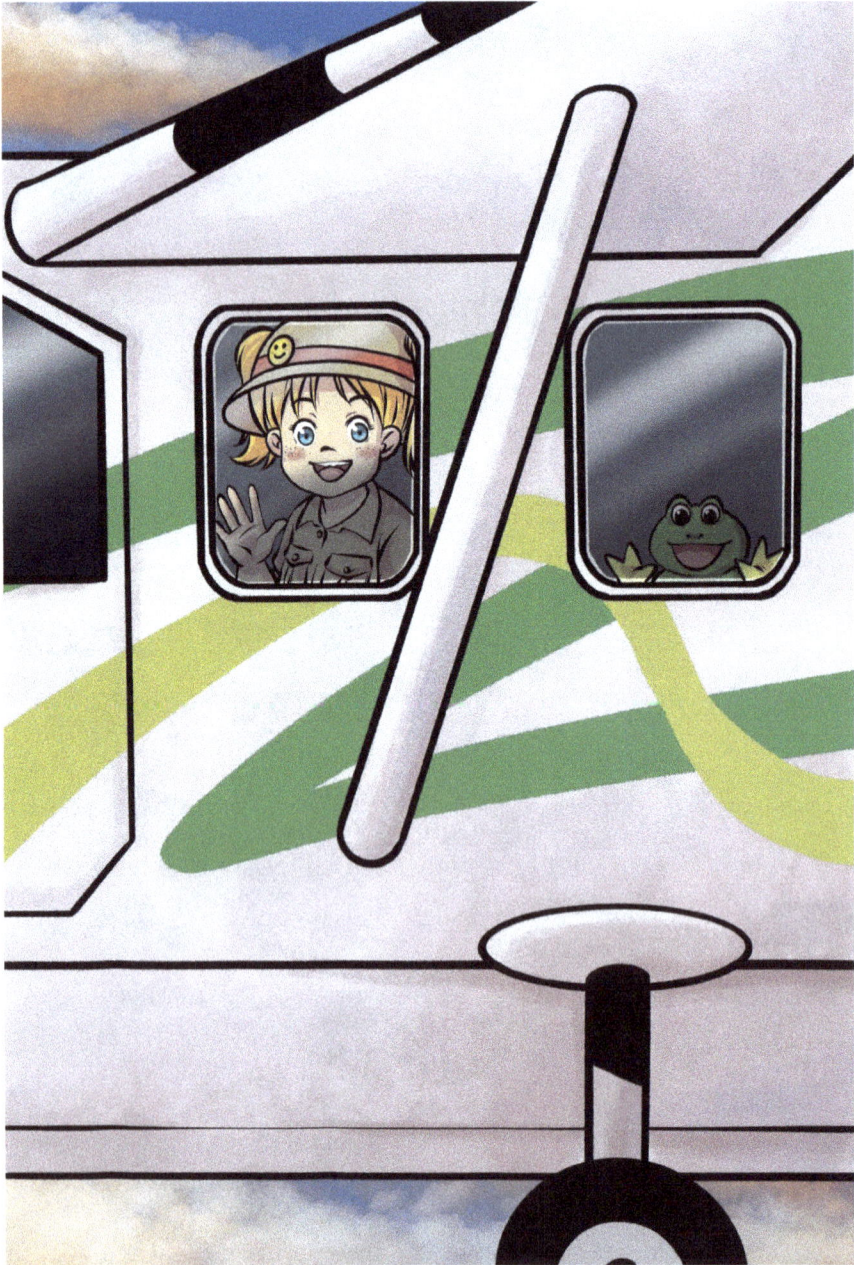

to a **protected park** in **Africa**

It's where the GREAT Gorillas live

Some people call gorillas gentle giants

The oldest boys are called Silverbacks - because

they have grey hair!

And guess what the teenage boys are called?

That's right!
- Blackbacks

Because they have black hair

Shhhh..

This Gorilla is beating his chest

His chest beating can be heard by gorilla families

WAY over on the other mountain

That's because his chest is hollow - like a **drum**

Bash and I are holding very still so this Silverback

knows that we mean him no harm

We also need to let him know that we are his friends

We do this by making very low noises

like we are clearing our throat

Visit Susie and her Animal Friends at
KidsKyngdom.com

I love gorilla grunts

Not **nuts!**

Gorilla Grunts!

Most of the food gorillas eat are fruit and tree bark

But this tribe of Silverback gorillas

are having a great big feast

That's right!
They are eating Ants!

Yum, Yum!

Let's join them, Bash!

You hop on this Mom's back

Hang on tight!

Visit Susie and her Animal Friends at
KidsKyngdom.com

Did you know that **baby gorillas** hang on to their

Mom's backs all day long!

That's right!

Until they are 4 years old!

Gorillas are also super smart!
They use tools
And can even speak sign
language with us

That is why I call them my cousins

What's that?

I hope our Kyngs can visit you too someday!

Hey Kyngs,
Gorillas live and play freely
ONLY in **Africa**

So let's be **careful** to **protect**
their home

Yeah!
Let's go VISIT!
I know another animal

that likes ants

That's right, Kyngs!

We're visiting an **Anteater**!!

Visit Susie and her Animal Friends at
KidsKyngdom.com

ISBN: 978-1-959501-06-0

Published by: Kyngdom, LLC

Visit Susie and her Animal Friends at
KidsKyngdom.com

What is Sammie Kyng's Favorite Animal?

Sammie **Kyng's** favorite animal used to be an **Gorilla.**

Now it's an **Anteater**!

If you have a favorite animal you would like Doctor Susie to visit, just **send a request** to Kidskyngdom.com or Instagram.

Kidskyngdom.Com   Instagram

Did you Love
learning about Gorillas
with Susie and Bash?

Please Share Your Review

Do You Know **oldest gorilla boys** are called?

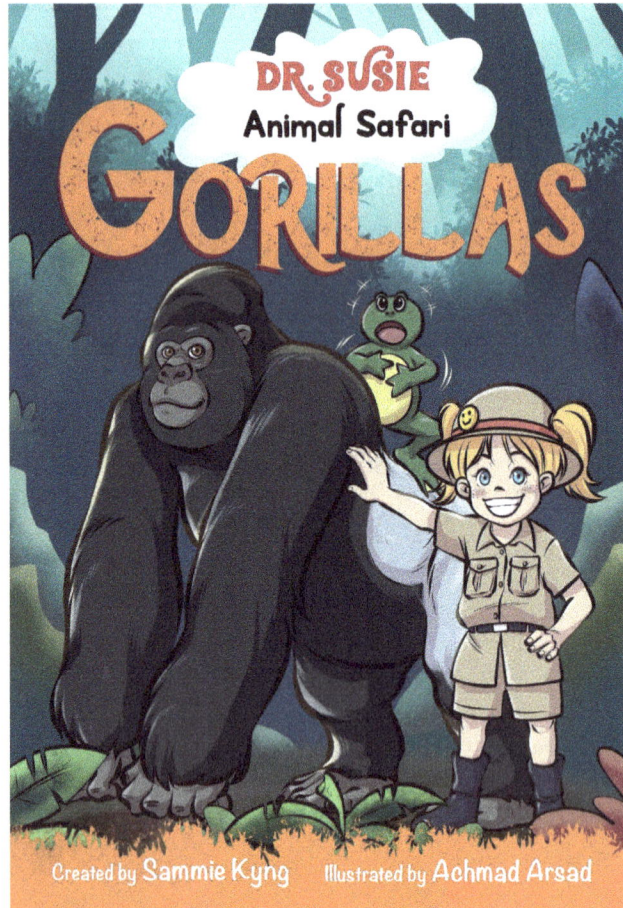

DR. SUSIE
Animal Safari
GORILLAS
Created by Sammie Kyng   Illustrated by Achmad Arsad

Search for Answers with Susie and Bash

Visit Kidskyngdom.com